Catherine Quévremont
Photographs by Sophie Tramier

Tomatoes

Styling: Garlone Bardel

HACHETTE
Illustrated

contents

Party finger food

Comfort food

Portrait of the tomato

The origins of this familiar vegetable lie in Peru, where the Aztecs, who also introduced the world to such edible delights as chocolate, gave it the name of *tomatl*. Imported into Spain in the 16th century, *Lycopersicon esculentum*, a member of the nightshade family, took about two hundred years to become accepted as edible. It was first considered an ornamental plant, and likely to be poisonous, but its looks earned it the name of *pomodoro* ('golden apple') in Italy. In North America, its cultivation didn't become widespread until the 19th century. Despite its difficult past, the tomato is now the most eaten vegetable in the world.

Today, it is difficult to imagine how dishes, from the kitchens of the most celebrated chefs to the sandwiches in our humble lunchbox, could exist without the variety of flavors and colors that the tomato provides. From the versatile bright red globe, to the yellow, green and even black Kumato varieties, the tomato has become an essential ingredient. Pasta without its scarlet sauce, burgers without a blob of ketchup, a BLT sandwich without the T, or a salad without that cheerful red ingredient—mealtimes would be much duller and less colorful without the Aztec gift.

Nothing beats the taste of a home-grown, sun-ripened and freshly picked tomato, as greenhouses crammed with tomato-laden plants testify. Even the apartment-dweller with a balcony that can accommodate a growing-bag can have a fresh tomato in season—between mid-summer and the first frost. If space is really limited, a hanging basket planted with Sun Cherry and Washington Cherry, or red and yellow Tumbler, provides a colorful and tasty cascade of cherry tomatoes during the summer months. Garden centers will have tomato plants all ready to go into patio containers, especially the bush varieties, which are sturdy plants producing good crops of small-sized fruits.

Refrigerated long-haul transport means that the tomato, once a summer season vegetable, is now available year-round, but in many cases at the cost of flavor. The winter months can find supermarket shelves piled with hard, pale ghosts, poor substitutes for the bright red, juicy globes of summer. It is often better to pay a little more for the vine tomato that has traveled attached to its stalk and has retained its color and a reasonably good flavor.

The right tomato for the job

Rich in vitamins A, B and C but low in calories, this essential ingredient is a firm favorite with trencherfolk and slimmers alike. The Italians were the first to spot the potential for the tomato cooked in sauces and have never looked back, but for the best results you should select the right type of tomato for the recipe. A rich, tasty bolognese sauce (see p. 8) needs a plum tomato with its firm flesh and low water content, a variety much loved by Italians and the mainstay of its canned tomato industry. Look out for San Marzano and Tuscany.

Beefsteak tomatoes, such as Brandywine or Big Beef, are ideal for sandwiches as the large slices have less chance of slipping out onto one's lap, a hazard with small sliced varieties. Although juicy they are not watery, so no more soggy sandwiches! Their firm flesh also makes them ideal for stuffing, providing a substantial meal (see p. 54). They are also delicious thinly sliced with mozzarella cheese and sprinkled with basil leaves and olive oil—a popular dish in Italy as the three indigenous ingredients also comprise the colors in their national flag.

As well as furnishing the country's salad bowls in its raw state, the familiar globe tomato will simmer down into delicious concentrated sauces for homemade pizza – or ketchup, to make a change from shaking and tapping that bottle. This type of tomato is also good for gazpachos, as are vine tomatoes.

The tiny cherry tomatoes are a delight in salads and hors d'oeuvres. Kept in their whole state, no juice escapes onto other ingredients, spoiling artistic and appetizing arrangements. At cocktail parties, sliced in half, their centers scooped out and filled with cream cheese and garnished with a dusting of mild chili powder or chopped parsley, this plate of dainty tidbits makes a good companion to the larger tomatoes stuffed with shrimp (see p. 46).

Skinning tomatoes

Many people's digestive systems are intolerant to tomato skins and, ascetically, many dishes are better without them. It is worthwhile to take a few minutes to do the job properly. Cut a small cross in the base of each tomato but do not cut into the flesh. Put them in a heatproof bowl, pour over boiling water, and leave for about 15–30 seconds. When the skins start to split, scoop them out and plunge into ice-cold water. The skins should come off easily but, if any stick, just pop them back into the hot water for about 10 seconds.

Preserved tomatoes

Ripe, medium-sized tomatoes, preferably home grown

sea salt, sugar, freshly ground black pepper

garlic cloves

olive oil

Preheat oven to 325°F.

Wash and cut the tomatoes in half, and remove the seeds with the tip of a knife.

Line a baking sheet with parchment paper and place tomatoes cut side up, side-by-side without crowding them.

Prepare a salt-pepper-sugar mixture, made from 1 tablespoon sea salt, $1/2$ teaspoon sugar and $1/2$ teaspoon freshly ground black pepper. Peel and crush the garlic. Season the tomatoes with the salt-pepper-sugar mixture and sprinkle them with the crushed garlic.

Bake for 2 hours, keeping an eye on the tomatoes as they cook; depending on how hot the oven is, they may require more or less time. The tomatoes are done when their juices have evaporated and they have shriveled up.

Leave to cool. Place in a sterilized and absolutely dry jar and cover with olive oil. Store in a dark place during the summer.

Note: In the Mediterranean, tomatoes are traditionally sun-dried.

Tomato coulis

4 lb large plum tomatoes

3 onions, peeled

1 celery stalk

1 thick slice cured ham

2 tablespoons olive oil

3 garlic cloves, crushed

thyme, bay leaf, savory

pinch of sugar, if the tomatoes are tart

salt and freshly ground black pepper

Skin the tomatoes (see p. 4) and cut into quarters. Thinly slice the onions and cut the celery into chunks. Cut the ham into thin strips.

Heat the oil in a large skillet or pan and add the onions and ham. Cook until the onions are soft. Add the tomatoes, celery, crushed garlic, thyme, bay leaf, savory, and sugar if needed. Cover and simmer gently for 1 hour.

Remove from the heat and allow to cool. Whizz in a blender or with a hand-held blender until smooth. Season to taste.

Pour into a glass jar and cover the surface of the coulis with a dash of olive oil (to keep it fresh). Keep refrigerated.

Bolognese sauce

2 onions

4 tablespoons olive oil

1 garlic clove, crushed

1½ cups raw ground beef

1¼ cups tomato coulis (see recipe above)

1¼ cups veal stock (or made from a cube)

½ cup freshly grated Parmesan cheese

sprig of thyme,

1 bay leaf

1 bunch of parsley, chopped

salt and freshly ground black pepper

Peel and thinly slice the onions.

Heat the oil in a large skillet. Throw in the onions and the crushed garlic, and cook until transparent. Add the ground beef, stirring often so that it browns well.

Pour in the tomato coulis and veal stock and season with salt and pepper. Dust with the Parmesan, crumble in the thyme leaves and bay leaf, and sprinkle with the chopped parsley. Cook gently for 35 minutes, stirring regularly.

When done, lightly process the sauce in a food processor or put through a mouli. Don't blend too smooth, as you want the meat to retain some of its texture.

Tomato sauce with broiled bell peppers

2 lb very ripe, globe tomatoes

2 red bell peppers

2 green bell peppers

⅔ cup olive oil

3 onions, peeled and thinly sliced

4 garlic cloves, crushed

½ bunch of cilantro, chopped

½ bunch of parsley, chopped

½ bunch of basil, leaves torn into small pieces

salt and freshly ground black pepper

Preheat the oven to 400°F.

Skin the tomatoes (see p. 4) and cut into quarters.

Place the whole bell peppers in a roasting pan and bake in the oven, turning them regularly until they are blackened and blistered all over. Remove from the oven and put them in a sealed plastic bag for 15 minutes, then peel and seed. Coarsely chop the flesh of the bell peppers.

Heat half the oil in a pan and sauté the onions, covered with a lid, until transparent.

Add the tomatoes to the pan. Pour in the remaining oil, and add the crushed garlic, chopped herbs, and bell peppers. Season with salt and pepper.

Simmer for 20 minutes. When ready, blend until smooth.

Homemade ketchup

4 lb very ripe globe tomatoes

1 lb red bell peppers

4 tablespoons olive oil

1 lb onions, thinly sliced

¹/₂ cup superfine sugar

1 cup white wine vinegar

1 teaspoon salt

2 tablespoons hot mustard

³/₄ cup water

¹/₂ teaspoon freshly grated nutmeg

paprika

chili powder, *Piment d'Espelette* if possible

cayenne pepper

Preheat the oven to 400°F.

Skin the tomatoes (see p. 4), seed and cut into quarters.

Place the whole bell peppers in a roasting pan and bake in the oven, turning them regularly until they are blackened and blistered all over. Remove from the oven and put them in a sealed plastic bag for 15 minutes, then peel.

Cut the skinned bell peppers in half, remove seeds and ribs, and cut the flesh into small dice, or process them coarsely.

Heat the olive oil in a large, high-sided skillet, then add all the vegetables and cook for 45 minutes, stirring frequently.

Blend in a food processer or mouli, and return the purée to the pan, adding the sugar, vinegar, salt, mustard, water, and spices to taste. Cook gently for 40 minutes, stir, check the thickness and add water if necessary. Adjust the seasoning to taste and pass through a conical strainer.

Note: This homemade ketchup may be kept in small sterilized bottles or jars. It can also be or stored frozen in plastic tubs or freezer bags. Keep in refrigerator after opening.

Spiced tomato chutney

4 lb just-ripe, globe tomatoes

3 onions

Fruit:

4 slightly tart apples

3 firm pears

1 cup golden raisins

juice of 1 lime

1¼ cups cider vinegar

⅔ cup brown sugar

1 teaspoon salt

Spices:

2 whole cloves

1 teaspoon mustard seed

1 cinnamon stick

few small, very hot chiles

1 teaspoon freshly grated ginger

Skin the tomatoes (see p. 4), then seed and crush them. Peel and thinly slice the onions. Peel the apples and pears and cut them into small dice.

Place all the fruit, tomatoes and onions in a large pan. Add the lime juice, vinegar, brown sugar, and salt, and stir together.

Tie up all the spices and chiles in a small piece of cheesecloth and place in the pan. Bring to a boil, lower the flame, cover with a lid and continue to simmer, stirring often. The chutney is ready once all excess liquid has evaporated. You should be left with a thick compote.

Ladle into dry, sterilized jars immediately and let mature for 1 month before using.

Tomato juice

2 lb well ripened globe
or plum tomatoes

1 celery stalk, diced

pinch of sugar

juice of 1 lime

few dashes of Tabasco

pinch of celery salt

salt and freshly ground
black pepper

Wash and cut the tomatoes into quarters. Place in a blender or juice extractor, together with the diced celery, the pinch of sugar, and the lime juice. Season with salt and blend until smooth.

For a little more punch, add 4 drops of Tabasco (to 1¾ pints of juice). Serve poured over ice, with celery salt, pepper and a dash of vodka.

Note: A fresh and natural aperitif that can be prepared in a flash and kept for several days in the refrigerator, but you can use canned tomato juice if you prefer.

Tomato granita

2 lb very ripe globe
tomatoes

pinch of sugar

juice of 1 lime

pinch of salt

freshly ground black pepper

a few dashes of Tabasco

Skin the tomatoes (see p. 4), seed and cut into quarters. Place in the blender jar with the sugar, lime juice, salt, pepper, and Tabasco to taste. Blend until smooth.

Pour the mixture into an ice-cream maker. During the chilling process, stir twice with a fork to break up the ice crystals and give it that granular appearance.

If you are keeping the granita in the freezer, let it thaw a little before serving.

Serve as an appetizer, as an accompaniment to fish in aspic, or as a first course followed by a tomato tart Tatin (see p. 32).

Tomato conserve

4 lb plum tomatoes

2 lb preserving sugar

4 tablespoons lemon juice

³/₄ cup water

1 vanilla bean

Skin the tomatoes (see p. 4), seed and chop coarsely.

Place the sugar and lemon juice into a preserving pan and add the water. Heat until the sugar dissolves. Once a syrup has formed, add the prepared tomatoes. Split the vanilla bean down the middle with the point of a knife; scrape out the tiny black seeds and add them to the tomatoes. Bring to the boil, then cook over a low heat for 45 minutes, until the tomatoes are translucent. Pour at once into dry, sterilized jars.

Tomato aspic

3lb ripe globe tomatoes, or use canned tomatoes

powdered gelatin

sprig of tarragon

Skin the tomatoes (see p. 4), seed and chop coarsely. Place in a large heavy-based pan and bring to the boil. Lower the heat and simmer for around 5 minutes. Let stand to to cool then filter through a colander or strainer, lined with cheesecloth.

Measure the juice and allow 1 tablespoon of powdered gelatin for every 2 cups collected. Heat the juice for 5–6 minutes in a pan, then add the tarragon. Allow to infuse for 15 minutes, then remove the tarragon.

Pour hot water into a cup and add the gelatin (never the other way round!). Stir to dissolve completely. Add the dissolved gelatin to the lukewarm tomato juice and stir well. Use as required, serving the final dish well chilled.

Note: Oysters in tomato aspic can be prepared with this recipe. Don't wait until the aspic is completely set before pouring it into the mold; use it when it has just begun to set. And don't waste the tomato peelings. Fry them with a few sage leaves in a little olive or peanut oil, and serve with a predinner drink.

Gazpacho

2–2½ lb very ripe globe, vine tomatoes

4 slices sandwich loaf

1 red bell pepper

1 green bell pepper

1 yellow bell pepper

3 garlic cloves

5 small fresh white onions

1 cucumber

2 tablespoons sherry vinegar

chili powder, preferably *Piment d'Espelette*

4 ice cubes

salt and freshly ground black pepper

4 tablespoons olive oil, to serve

Skin the tomatoes (see p. 4), seed and cut into quarters.

Remove the crust from the bread, then soak in 1 cup of water (or vegetable stock).

Cut the bell peppers in half, remove their seeds and ribs and cut into small dice. Peel the garlic, removing the green shoot in the middle, and slice finely. Peel and thinly slice the onions. Peel the cucumber and cut into small dice.

Squeeze out the bread and place in a blender with the tomatoes, garlic, and vinegar. Blend until smooth. Season with salt and pepper and sprinkle with the chili powder, to taste. Pour into a large bowl and add the ice cubes. Chill until ready to serve.

Serve the gazpacho accompanied by all the diced vegetables. Sprinkle with olive oil just before serving.

Cream of tomato soup with sage

4 floury potatoes, peeled

6 large, ripe plum tomatoes

1 onion, peeled

2 garlic cloves, peeled

2 tablespoons olive oil

pinch of sugar

16 sage leaves

$2/3$ cup whipping cream

salt and freshly ground black pepper

Cut the peeled potatoes into large cubes and set aside in a bowl of cold water.

Skin the tomatoes (see p. 4), seed and chop coarsely. Thinly slice the onion and garlic.

Place half the olive oil in a pan and sweat the onion and garlic in the hot oil. Add the tomatoes, season with salt and pepper and sprinkle over the pinch of sugar. Simmer for 5 minutes.

Drain the potatoes and add to the pan, together with $1^3/4$ pints water and half the sage leaves. Simmer over a moderate heat for 30 minutes or until the vegetables are tender.

Blend until smooth. Check the seasoning.

Fry the remaining sage leaves in the rest of the olive oil. Pour the soup into a tureen, add the whipping cream and stir just until it leaves a creamy white trail on the surface, then sprinkle with the fried sage leaves.

Moroccan salad

1 red bell pepper

1 green bell pepper

4 ripe salad tomatoes

1 bunch of parsley, finely chopped

3 tablespoons lemon juice

3 tablespoons olive oil

1 teaspoon ground cumin

salt

12 large green or black olives, to garnish

Preheat the oven to 400°F.

Place the whole bell peppers in a roasting pan and bake in the oven for 30 minutes, turning them regularly until they are blackened and blistered all over.

While the peppers are baking, skin the tomatoes (see p. 4), seed and cut into small dice.

Remove the bell peppers from the oven, seal in a plastic bag and leave to cool. They will now peel very easily. Seed and cut the flesh into small dice.

Combine the tomatoes, bell peppers, and chopped parsley. Dress with the lemon juice and olive oil, and sprinkle with the ground cumin and salt.

Serve on individual plates, garnished with large green or black olives.

Tomato and mozzarella salad with pesto

6 tomatoes

3 balls of *mozzarella di bufala* (buffalo milk mozzarella)

1 bunch of basil

½ cup pine nuts

½ cup freshly grated Parmesan cheese

pinch of coarse salt

2 garlic cloves, peeled

6 tablespoons olive oil

2 tablespoons dry white wine

salt and freshly ground black pepper

Wash the tomatoes and remove their stalks. Cut into thin, even slices.

Open the packets of mozzarella, drain the water and pat dry with paper towels. Slice the mozzarella and arrange the slices on a platter, alternating them with the sliced tomatoes. Season with salt and pepper.

Wash and dry the basil and strip the leaves from the stalks, reserving a few for garnish. Place the basil leaves, pine nuts, Parmesan, coarse salt, and peeled garlic in a blender. Blend, adding the olive oil in a steady stream. This will produce a thick pesto that can be diluted with the white wine to form a dressing.

Pour the pesto dressing over the tomatoes and mozzarella and garnish with a few basil leaves.

Tomato conserve tart

15 cherry tomatoes

¼ cup/½ stick butter

4 tablespoons Demerera sugar

1-inch piece fresh root ginger, peeled

8 oz plain pastry

12-oz jar tomato conserve (see p. 18)

20 plain pistachios

Cut the tomatoes in half. Melt the butter in a large skillet and quickly brown the tomatoes in the butter. Sprinkle with sugar, then grate the ginger over the top. Caramelize the tomatoes, cooking them over a high heat and taking care to preserve their appearance as far as possible.

Preheat the oven to 400°F.

Roll out the pastry and use to line a tart dish, or 6 individual tart tins. Spread the tomato conserve over the base of the pastry and arrange the caramelized tomato halves on top.

Bake in the preheated oven for 30 minutes. Crush the pistachios and sprinkle over the tart as soon as it is removed from the oven.

Tomato and mustard tart

6 plum tomatoes

8 oz plain pastry

3 tablespoons Dijon mustard

³/₄ cup freshly grated Parmesan cheese

½ teaspoon dried oregano

salt and freshly ground black pepper

Preheat the oven to 375°F.

Skin the tomatoes (see p. 4) and seed them. Cut the flesh into thin slices and place them on paper towels while you prepare the remaining ingredients.

Roll out the pastry and use to line a pie pan. Prick the base with a fork, then bake for 10 minutes in the preheated oven until the base is dry.

Remove the pastry shell from the oven, spread the mustard over the base and sprinkle with half of the grated Parmesan.

Turn the oven up to 400°F.

Arrange half of the tomato slices over the base; season with salt and pepper and sprinkle with oregano, then top with a second layer of tomato slices and oregano. Finish by sprinkling with the remaining grated Parmesan.

Bake for 20 minutes.

Tomato tart Tatin

2 lb plum tomatoes

1 bunch of basil

2 tablespoons olive oil

2 garlic cloves, thinly sliced

1 ball *mozzarella di bufala* (buffalo milk mozzarella)

few pinches of oregano

8 oz ready-made puff pastry

salt and freshly ground black pepper

Skin the tomatoes (see p. 4), seed and crush. Remove the leaves from the basil and shred finely.

Heat the oil in a large skillet and add the garlic, crushed tomatoes, and basil. Season with salt and pepper and cook over high heat until you have a slightly moist purée.

Preheat the oven to 400°F.

Open the packet of mozzarella, drain off the water and pat the cheese dry with paper towels. Cut into thick slices. Arrange the cheese slices in the base of a greased pie pan, then spread the tomato mixture on top. Sprinkle with oregano. Roll out the puff pastry to fit your pie pan and place it on top of the tomato mixture, tucking in the edges to seal.

Bake the tart Tatin in the preheated oven for 25 minutes. Allow to cool slightly and then unmold the tart by carefully turning it upside down onto a serving plate.

Tomato bruschetta

6 plum tomatoes

6 tablespoons olive oil

1 bunch of basil

6 large slices from a
country-style loaf

3 garlic cloves, peeled

3½ oz arugula

vinaigrette

salt and freshly ground
black pepper

Skin the tomatoes (see p. 4), seed and cut into small dice.
Mix with 2 tablespoons of the olive oil and season with salt
and pepper.

Remove the leaves from the basil and finely shred.

Brush the slices of bread with the remaining olive oil and toast
under the broiler. Rub the peeled garlic over the bread.

Spoon the diced tomato onto the bread, and garnish with the
finely shredded basil leaves.

To serve, cut the bruschettas into 1½-inch squares and
serve on a bed of arugula leaves, dressed with a little
vinaigrette to taste.

Pizza napolitana

5 plum tomatoes
3 garlic cloves, peeled
1 fresh or frozen pizza crust
4 tablespoons olive oil
1 tablespoon dried oregano
salt

Preheat the oven to 400°F.

Skin the tomatoes (see p. 4), seed and crush. Finely slice the garlic. Brush the pizza crust with a little of the olive oil.

Mix the crushed tomatoes, garlic and oregano in a bowl. Season with salt. Spread this mixture over the pizza base.

Bake for a good 10 minutes.

Remove from the oven and drizzle with the remaining olive oil.

Variation: To make a Margherita version of this pizza, add thin slices of mozzarella cheese 5 minutes before the pizza is done, and garnish with chopped basil before serving.

Anglerfish and tomato terrine

2 lb globe tomatoes

2 tablespoons olive oil

6 slices from a white sandwich loaf

3 tablespoons milk

2 lb skinned anglerfish

1 small can tomato paste

6 eggs

¾ cup heavy cream

1 tablespoon lemon juice

1 bunch of tarragon, leaves removed

salt and freshly ground black pepper

Skin the tomatoes (see p. 4), seed and crush.

Heat the olive oil in a skillet and throw in the tomatoes. Season with salt and pepper and simmer until reduced to a very dry purée.

Remove the crusts from the slices of bread, crumble the bread into a bowl and pour over the milk. Leave to soak for a few minutes before squeezing out.

Cut half of the anglerfish into 1-inch cubes. Chop the rest of the anglerfish and place in a blender with the tomato purée, the can of tomato paste, eggs, squeezed-out bread, heavy cream, lemon juice and half of the tarragon leaves. Taste and correct seasoning with salt and pepper.

Preheat the oven to 350°F.

Pour some of the mixture into a terrine; arrange the anglerfish cubes and a few tarragon leaves on top. Pour over the rest of the mixture, finishing with some tarragon leaves. Place the terrine in a roasting pan and pour in hot water to half way up the terrine. Bake in the preheated oven for 1 hour.

Remove from the oven, allow to cool then place in the refrigerator overnight. The terrine should be eaten cold, accompanied by a tomato coulis (see p. 8), or cut into cubes, skewered onto toothpicks and served with preserved tomatoes (see p. 6).

Goats' cheese and tomato terrine

8 round tomatoes

2 white onions, peeled

2 garlic cloves, peeled

3 tablespoons olive oil

5 gelatin leaves

2 fresh goats' cheeses

2 tablespoons heavy cream

lemon juice

paprika

salt and freshly ground
black pepper

Skin the tomatoes (see p. 4) and seed. Thinly slice the onions and garlic.

Heat the olive oil and gently simmer the tomatoes, onions, and garlic until you have a dry purée. Season to taste with salt and pepper.

Soak the gelatin leaves, separating them into one group of 2 and one of 3. Squeeze out the 2 gelatin leaves, then stir them into the tomato mixture.

Crush the goats' cheeses with a fork. Gently heat the heavy cream and add the 3 reserved gelatin leaves, which have been soaked and squeezed dry. Mix thoroughly and pour over the goats' cheese, adding a little lemon juice to taste. Mix well to yield a creamy paste. Dust with paprika.

Line a terrine mold with plastic wrap leaving a generous 'overhang'. Pour in a layer of the goats' cheese mixture, then top with a layer of the tomato mixture, continuing in this manner until both mixtures are used up. Fold over the plastic wrap to seal, tamp down slightly and refrigerate overnight.

Tomato and salt cod mold

1½ lb salt cod fillets

5 garlic cloves, peeled

¾ cup olive oil

1¼ cups very dry tomato purée from 2 lb tomatoes, see anglerfish and tomato terrine (see p. 38)

¾ cup preserved tomatoes (see p. 6)

savory and thyme

Soak the cod fillets for 24 hours to remove the salt, changing the water several times. The next day, poach the cod for 10 minutes in plenty of unsalted water. Drain and skin the fish, remove any bones and flake.

Place the cod flakes and the garlic in a blender and process to obtain a smooth purée, adding the olive oil in a slow, steady stream.

Clean the blender bowl, then coarsely blend the tomato purée and the preserved tomatoes. Avoid overblending; the mixture should retain some texture.

Line a terrine mold with plastic wrap leaving a generous 'overhang'. Pour a layer of the tomato mixture in the bottom of the mold, followed by a layer of the cod mixture, and so on until all the ingredients are used up.

Crumble the savory and thyme over the top. Fold over the plastic wrap and seal. Weight the terrine down (with a 1 lb can) and refrigerate overnight, or for at least 6 hours.

To serve, open the plastic wrap, unmold the terrine, and cut into slices with an electric carving knife (the terrine is fragile) or a very sharp knife, dipped in hot water. Serve with a spoon if preferred.

Green tomato tempura

6 green tomatoes

1 cup peanut oil

1 cup sesame oil

salt

sesame seeds

For the tempura batter:

1 egg

4 tablespoons sifted all-purpose flour

½ small bottle of chilled sparkling water

Wash the tomatoes and remove their stalks. Slice thinly and seed, then set aside on paper towels while you prepare the tempura batter.

Separate the egg and stiffly beat the white. Mix the flour with the egg yolk, then beat in sufficient of the chilled sparkling water to make a thin batter. Beat until the batter is smooth, then fold in the stiffly beaten egg white.

Heat both oils together in a large, high-sided skillet. Dip each tomato slice in batter and fry in the hot oil. As soon as the batter begins to swell up on one side, turn over the fritter and finish cooking it on the other side.

Place the tomato fritters on paper towels to drain as you finish frying them. Season with a little salt and sprinkle with sesame seeds.

Tomatoes stuffed with shrimp

12 globe tomatoes

1 bouquet garni

1 onion, peeled

1 clove

1 lb live tiny shrimp

1 cup freshly made mayonnaise, or best quality store bought

coarse salt and freshly ground black pepper

Slice off the tops of the tomatoes and scoop out their contents with a small spoon. Salt the insides and place them upside-down on paper towels for 30 minutes to allow the natural juices to drain.

Fill a large pan with water. Throw in a handful of coarse salt, and add 2 or 3 turns of the pepper mill, the bouquet garni, and the onion studded with the clove. Bring to the boil, add the shrimp to this court-bouillon, and bring back to the boil. Turn off the heat, cover the pan, and leave undisturbed for 5 minutes.

Drain and peel the shrimp. Mix them with the mayonnaise and stuff the tomatoes with this mixture. Chill until you are ready to serve.

Cream cheese and prawn stuffing

16 shrimp

7 oz Boursin cheese with mixed herbs

Peel the shrimp and cut into 2 or 3 pieces. Crush the Boursin with a fork, mix with the shrimp, and stuff the tomatoes with this mixture.

Baby squid in tomato sauce

3 lb baby squid

2 lb ripe globe tomatoes

3 garlic cloves, peeled

3 onions, peeled

1 bunch of parsley

½ bunch of cilantro

4 tablespoons olive oil

1 teaspoon chili powder, preferably *Piment d'Espelette*

salt and freshly ground black pepper

Prepare the squid, or have it done for you. Cut the tentacles level with the head, slice open the body, remove the quill and clean the body cavity. Rinse the prepared squid several times. Cut the bodies into thin rings and arrange them on paper towels to drain.

Skin the tomatoes (see p. 4), seed and crush. Thinly slice the garlic and onions. Finely chop the parsley and cilantro.

Heat the olive oil in a large, high-sided skillet. Sweat the onion and add the garlic. Once the onions turn translucent, add the tomatoes.

Stir in the pieces of squid, season with salt and pepper and sprinkle with chili powder. Stir gently and simmer for 30 minutes. Ten minutes before the end of the cooking time add the parsley and cilantro.

Once the sauce has thickened, it's ready.

Osso buco

4 globe tomatoes

2 carrots, peeled

1 celery stalk

1 onion, peeled

1 garlic clove

1/4 cup/1 1/2 stick butter

1 tablespoon olive oil

6 slices knuckle of veal,
1-inch thick

1 cup white wine

1 cup veal stock (made from
a stock cube)

1 bouquet garni

4 tablespoons lemon juice

grated rind of 1 lemon

salt and freshly ground
black pepper

2 tarragon sprigs, to garnish

Skin the tomatoes (see p. 4), seed and cut into large pieces. Dice the carrots and celery. Thinly slice the onion and garlic.

Heat the butter and oil in a heavy, heatproof casserole. Add the pieces of veal knuckle and brown them with the garlic and onion, turning the meat several times so that it is golden on all sides. Add the wine and bring to the boil to evaporate the alcohol. Pour in the veal stock.

Add the vegetables, bouquet garni, lemon juice, and grated lemon rind. Simmer, covered, for 1 1/2 hours.

Serve the osso buco in a large dish garnished with tarragon sprigs.

Note: The osso buco is delicious served with fresh pasta drizzled with homemade tomato sauce.

Sautéed chicken with preserved tomatoes

3 garlic cloves

3 shallots

1/3 cup/3/4 stick butter

1 chicken cut into pieces, or 6 thigh joints

1 lb waxy potatoes, peeled

3/4 cup preserved tomatoes (see p. 6)

1 cup homemade ketchup (see p. 12)

1 cup dry white wine

1 bunch of basil, finely chopped

salt and freshly ground black pepper

Peel and thinly slice the garlic and shallots. Melt the butter in a heavy heatproof casserole with a lid and sweat the garlic and shallots, then add the chicken pieces and brown them well on all sides. Remove the chicken pieces from the pot and set aside.

Cut the potatoes into thick rounds and place them on the bottom of the casserole. Arrange the chicken pieces on top of the potatoes. Coarsely chop the preserved tomatoes and scatter over the chicken. Pour in the ketchup and white wine, and sprinkle in half the basil. Season with salt and pepper.

Cover with the lid and cook over a low heat for 1 hour (the juices of the chicken should run clear when tested with a sharp skewer). Check the sauce 10 minutes before the end of the cooking time; if it seems too thin, remove the lid, turn up the heat a little and reduce to the preferred consistency.

To serve, arrange the chicken and vegetables on a serving plate, and garnish with the remaining finely chopped basil.

Tomatoes stuffed with meat

12 medium or 6 large
beefsteak tomatoes

3 garlic cloves, peeled

3 shallots, peeled

1 onion, peeled

3 tablespoons olive or
peanut oil

1 lb sausagemeat

10 oz chopped cooked beef
(leftovers from a roast or
joint)

½ bunch of cilantro,
coarsely chopped

1 bunch of chives, coarsely
chopped

four-spice powder
(see note below)

ground cumin

salt and freshly ground
black pepper

Preheat the oven to 400°F.

Slice the tops off the tomatoes and set aside as "lids". Scoop out the insides from each tomato with a small spoon, discarding the seeds and core. Salt the insides and place upside-down on paper towels for about 20 minutes to drain. Chop the flesh into dice.

Finely chop the garlic, shallots and onion. Heat the oil in a large pan and cook the garlic, shallots, and onion until golden. Add the diced tomatoes and crumble the sausagemeat into the pan. Brown the sausagemeat, stirring, then add the cooked beef. Stir so that it cooks evenly. Quite a lot of moisture should evaporate from the mixture.

Wash the fresh herbs. Place the meat mixture and herbs in a food processor (you may need to do this in several stages). Season to taste with salt and pepper, add the four-spice powder and ground cumin, also to taste, and process well. Stuff the tomatoes with the meat mixture and top with the "lids". Arrange the tomatoes in a large ovenproof dish and bake for 40 minutes.

Note: Four-spice powder (or Quatre-épices) is a mixture usually made from ground pepper, cloves and cinnamon, and grated nutmeg. It is available in good supermarkets.

Provençale tomatoes

12 globe tomatoes

6 garlic cloves, peeled

3 slices stale bread, grated (remove crusts first)

1 bunch of flat-leaf parsley, washed

2/3 cup olive oil

salt and freshly ground black pepper

Preheat the oven to 375°F.

Wash and halve the tomatoes. Gently squeeze each half to remove the seeds and some of the watery juices.

Tightly pack the tomatoes cut-side up in a large ovenproof dish. Season with salt and pepper.

Process the breadcrumbs, garlic, and parsley in a blender, and mix with half the olive oil.

Spread this paste over the tomatoes. Drizzle with the remaining olive oil and bake in the preheated oven for 20 minutes.

Risotto with preserved tomatoes

1 onion, peeled

1 garlic clove, peeled

2/3 cup olive oil

1 cup tomato concassée
(skinned, seeded and diced
tomatoes)

1 lb risotto rice (arborio
or carnaroli)

1¾ pints chicken stock
(made from a cube)

20 preserved tomatoes
(see p. 6)

salt and freshly ground
black pepper

5 oz fresh Parmesan in
a piece, to serve

Thinly slice the onion and garlic. Heat one-third of the oil in a pan and sauté the onion and garlic. Remove from the heat as soon as the garlic has colored slightly. Add the tomato concassé and allow to reduce at a very slow simmer for 20 minutes, until you have a thick purée.

Heat the remaining olive oil in a high-sided skillet or pan. Add the rice and stir until all the grains are thoroughly coated in oil. Once the rice turns translucent, add 1 ladleful of chicken stock and stir. As soon as the rice has absorbed all the liquid, add another ladleful and when it has been absorbed add another and so on until all the liquid has been used up and the rice is creamy, but still with a 'bite'.

Add the tomato sauce during the course of cooking and season to taste with salt and pepper.

Cut the preserved tomatoes into thin slices. Once the rice is cooked, add the tomatoes and mix well.

To serve, transfer the risotto to a large heated bowl and, using a paring knife or cheese shaver, cut the Parmesan into shavings and scatter over the risotto.

Ratatouille

2 lb plum tomatoes

2 lb zucchini

2 eggplants

1 red bell pepper

1 green bell pepper

1 cup peanut oil

2 onions, peeled and sliced

4 garlic cloves, crushed

sprig of thyme, bay leaf

pinch of sugar

4 tablespoons olive oil

salt and freshly ground black pepper

Preheat the oven to 375°F.

Skin the tomatoes (see p. 4), seed, chop coarsely and set aside. Wash the zucchini, eggplants, and bell peppers. Place the bell peppers in a roasting pan and bake in the preheated oven, turning regularly, until blackened and blistered all over. Remove from the oven and place in a sealed plastic bag for 15 minutes, then peel, seed, and cut the flesh into strips

Heat some of the peanut oil in a skillet or pan and sweat the onions. Season with salt. Once they have softened and taken on some color, drain on paper towels.

Add more oil to the pan and fry the strips of bell pepper to soften; season with salt and drain on paper towels. Cut the zucchini into thick rounds and fry, adding more oil if necessary. Season with salt, then drain in a sieve or colander. Cut the eggplants into large cubes, proceed in the same manner and drain. Finally, sauté the tomatoes with the crushed garlic, thyme, and bay leaf. Season with salt and pepper, then add a pinch of sugar. Cook until the watery juices have evaporated.

Combine all the separate components of the dish in a large pan, then cover and cook for another 30 minutes over a low heat. Remove the sprig of thyme and the bay leaf. Pour the olive oil over the hot ratatouille and stir to combine.

Note: "Genuine" ratatouille requires each ingredient to be cooked separately. For a shortcut, cut all the vegetables into tiny dice, throw them into a skillet with olive oil, and sauté over a high flame. The vegetables should remain crisp. This dish can be served with broiled meat, or on a bed of fresh pasta. It can also be used to stuff small tomatoes for a buffet.

Country-style vegetable sauté with sage

2 lb globe, vine tomatoes

3 lb eggplants

2 onions, peeled

4 tablespoons olive oil

4 garlic cloves, peeled

8 sage leaves

salt and freshly ground black pepper

Skin the tomatoes (see p. 4), seed, chop and crush.

Wash the eggplants, then top and tail them and cut into large dice. Thinly slice the onions and sauté in half the olive oil in a large, high-sided skillet. Once they have taken on some color, add the eggplant and continue to sauté.

Crush the garlic in a garlic press. Add the tomatoes and garlic to the skillet. Season with salt and pepper, scatter 4 sage leaves over the mixture, then cover and cook over a low heat for 30 minutes.

Fry the remaining sage leaves in a small pan in the rest of the olive oil and scatter them over the vegetable sauté just before serving.

Note: A few anchovy filets, carefully rinsed to remove excess salt and crushed, could also be added to this vegetable sauté, in which case less salt should be added to the mixture.

All the recipes in the original edition of this book were prepared with Saveol tomatoes. The author and the entire Marabout team would like to thank Saveol for the use of their produce, as well as for their valuable culinary advice.

Acknowledgments

The editor would like to thank the outlets for the loan of cutlery, serving bowls and equipment for the photographs:

Agapé: 91 Avenue Jean-Baptiste Clément, 92100 Boulogne; Allison Grant: 33 Rue du Poitou, 75003 Paris

Antoine et Lili: 51 Rue des Francs-Bourgeois, 75004 Paris; Astier de Villatte: 173 Rue Saint-Honoré, 75001 Paris

Au petit Bonheur la Chance: 13 Rue Saint-Paul, 75004 Paris; Caravane: 6 Rue Pavée, 75004 Paris

CSAO: 1-3 Rue Elzévir, 75003 Paris; Cuisinophile: 28 Rue du Bourg Tibourg, 75004 Paris

Galerie Sentou: 18 Rue du Pont Louis-Philippe, 75004 Paris; Jeannine Cros: 11 Rue d'Assas, 75007 Paris

Lou Delamare: 10 Rue de Montreuil, 94300 Vincennes; Luka Luna: 77 Rue de la Verrerie, 75004 Paris

L'Objet qui Parle: 86 Rue des Martyrs, 75018 Paris; Paris Happy Home: 76 Rue François-Miron, 75004 Paris

Les Toiles du Soleil: 66260 Saint-Laurent-de-Cerdans; Totor et Cie: 69 Rue de la Fontaine-au-Roi, 75011 Paris

Editor: Elisabeth Darets
Art direction: Emmanuel Le Vallois
Editorial: Rosemarie Di Domenico
Production: Laurence Ledru
Proofreading: Antoine Pinchot and Véronique Dussidour